安仁赶分社

"人类非物质文化遗产代表作——二十四节气"科普丛书

春雨惊春清谷天
夏满芒夏暑相连
秋处露秋寒霜降
冬雪雪冬小大寒

中国农业博物馆 组编

Popular Science Book Series on
A Masterpiece of the Intangible Cultural Heritage of Humanity — The Twenty-Four Solar Terms

Spring Equinox Fair in Anren

Compiled and Edited by China Agricultural Museum

中国农业出版社
China Agriculture Press

北京 Beijing

序

Foreword

　　中华文化，博大精深，灿若星河，传承有序，绵延不绝。作为人类非物质文化遗产代表、凝结中华文明智慧的"二十四节气"在我国自创立以来，已经传承发展 2 000 多年。它是中国人观天察地、认知自然所创造发明出的时间知识体系，也是安排农业生产、协调农事活动的基本遵循，更是中国社会顺天应时、指导实践的生活制度。它是中华优秀传统文化中文明成果的典型代表，体现了传统农耕文明的智慧性，彰显了中国人认知宇宙和自然的独特性及其实践活动的丰富性，凸现了中国人与自然和谐相处的哲学思想、文化精神和智慧创造。

Chinese culture, wide-embracing and profound, brilliant in numerous fields, inherited in an orderly manner, has been developing without interruption. The "Twenty-Four Solar Terms" is a masterpiece of the intangible heritage of humanity and a crystallization of Chinese civilization and wisdom for 2,000 years of its existence. It is a time knowledge system invented by the Chinese people to observe heaven and earth and to learn about nature, the basic principles to organize agricultural production and coordinate farming activities, and also the life system for the Chinese society in its conformation to natural and meteorological timings and in its guidance of daily practices. As the typical representation of the fruit of the best of the traditional Chinese civilization, it embodies the wisdom of traditional farming civilization, reflects the Chinese people's unique interpretation of the universe and nature and the rich practices therein, and highlights their philosophical concepts, cultural spirit and intelligent creativity in their harmonious with nature.

"二十四节气"起源于战国时期，在公元前140年就已经有完整的"二十四节气"记载。从时间上，作为太阳历，早于儒略历（公元前45年）近一个世纪。"二十四节气"较之公历更准确地标识了地球视角的太阳运行规律。农谚就"二十四节气"同公历的关系说道："上半年是六廿一，下半年来八廿三。每月两节日期定，最多不差一两天。"这里所说的"不差"，不是"二十四节气"不准，而是公历有"差"。我们的生活要顺天应时，生活在自然体系之中，就应该把自己看成是包括自然界在内的客观世界的组成部分。无限制地扩大人的能力，破坏自然规律，其后果是难以意料的。

The "Twenty-Four Solar Terms" originated in the Warring States Period. There were already complete records of the "Twenty-Four Solar Terms" in 140 B.C., almost one century earlier than the Gregorian calendar (45 B.C.), also a solar calendar. It is more accurate than the latter in indicating the laws of the sun's movement from the earth's perspective. Agricultural proverbs identify its relationship with the Gregorian calendar: "In the first half of the year, the solar terms fall on the 6th and 21st of each month, and in the second half, they fall on the 8th and 23rd. There are two solar terms in each month, with an adjustment of one or two days." Here, "adjustment" is not the result of inaccuracy of the "Twenty-Four Solar Terms" but errors on the part of the Gregorian calendar. As we need to conform to the natural and meteorological laws and live in natural systems, we should regard ourselves as components of the objective world, including the nature. Expanding human capacity without constraint and disrupting the natural laws may lead to unexpected consequences.

对中国人来说，"二十四节气"是我们时间制度整体的一部分，它是指导我们包括农业在内的创造生活资料的一切活动的时间节律。而我们的情感表达、礼仪等调节人际关系、社会关系的活动则以对月亮运行周期观察为基础的太阴历为节律。我们的传统节日体系大都是以太阴历为依据的。在我们的阴阳合历的整体框架里认识"二十四节气"，领会我们的先辈以置闰的办法精妙恰当地协调二者的对应关系，体现了中华传统文化的精奥和人文精神。

For the Chinese, the "Twenty-Four Solar Terms" is part of our time regime, the time prosody that directs all activities that produce living materials, including agriculture. The time prosody

④

“人类非物质文化遗产代表作——二十四节气”科普丛书
Popular Science Book Series on
A Masterpiece of the Intangible Cultural Heritage of Humanity — The Twenty-Four Solar Terms

of activities that accommodate interpersonal and social relations, such as our emotional expression and etiquette and protocol, is the lunar calendar based on observations of the moon's periodic movement. Our system of traditional festivals is mostly founded on the lunar calendar. It is advisable to interpret the "Twenty-Four Solar Terms" in the overall framework of the lunisolar calendar, and understand how our ancestors wisely and aptly coordinate the correspondence between the solar and lunar calendars by means of intercalation, which reflects the beauty and humanistic spirit of traditional Chinese culture.

“二十四节气”融合四季，贯穿全年，广为实践，流布全国，影响世界。其作为我国优秀传统文化的典型代表和人类非物质文化遗产代表作项目，富含中国人特有的哲学思想、思维理念和人文精神，具有广泛的参与度和社会影响力，引发世人的关注与探索。2016 年 11 月 30 日，在文化部非物质文化遗产司指导下，在中国民俗学会支持下，由中国农业博物馆作为牵头单位，联合相关社区单位申报的“二十四节气——中国人通过观察太阳周年运动而形成的时间知识体系及其实践”，被联合国教科文组织列入人类非物质文化遗产代表作名录。这是中国非遗保护工作取得的一项重要成果，也是对外文化交流的一次成功实践。在其带动影响下，全国人民乃至世界人民对“二十四节气”的认知、认同、参与和实践空前提高，进一步彰显和增强了中国人的文化自觉和文化自信。

The "Twenty-Four Solar Terms", integrating the four seasons and covering the whole year, is widely practiced throughout China, with influences on the whole world. As China's best

⑤

"人类非物质文化遗产代表作-----二十四节气"科普丛书
Popular Science Book Series on
A Masterpiece of the Intangible Cultural Heritage of Humanity — The Twenty-Four Solar Terms

representative of traditional culture and a masterpiece of the intangible heritage of humanity, it is full of philosophical thoughts, thinking patterns and humanistic spirit unique to the Chinese, enjoying a wide participation and social influence, and commanding attention and exploration from around the world. On Nov. 30, 2016, under the guidance of the Intangible Cultural Heritage Department of the Ministry of Culture and with the support of the China Folklore Society, "the Twenty-Four Solar Terms, knowledge of time and practices developed in China through observation of the sun's annual motion" submitted by the China Agricultural Museum together with related community organizations, was entered onto the list of Masterpieces of the Intangible Cultural Heritage of Humanity by the UNESCO. This was a significant achievement in China's intangible heritage protection, and also a successful practice in cultural exchange. Due to this endeavor and its influence, the people of China and of the world have unprecedentedly heightened their knowledge, identification, participation and practice regarding the "Twenty-Four Solar Terms", which further reflects and enhances Chinese people's cultural awareness and self-confidence.

出版这套"人类非物质文化遗产代表作——二十四节气"科普丛书，有助于在更大更广的范围和层面普及传播节气的相关知识，进一步增强遗产实践社区和群众的自豪感与凝聚力，激发传承保护的自觉性和积极性，扩大关于传统时间知识体系的国际交流与对话，推动人类文明交流互鉴。

The publication of this book series on *A Masterpiece of the Intangible Cultural Heritage of Humanity—The Twenty-Four Solar Terms* shall be conducive to its spread and popularization on a larger scale and in a wider sphere, further enhance the sense of pride and solidarity on the

6

"人类非物质文化遗产代表作——二十四节气" 科普丛书
Popular Science Book Series on
A Masterpiece of the Intangible Cultural Heritage of Humanity — The Twenty-Four Solar Terms

part of the inheritance practice communities and masses, inspire their awareness and initiative in preservation and protection, expand international exchanges and dialogues on traditional time knowledge systems, and promote exchanges and mutual learning between human civilizations.

期望并相信这套丛书能够得到社会各界人士的喜爱。

We sincerely hope and cordially believe that this series will win the hearts of readers of various circles.

谨为序。

Please enjoy your reading of this volume.

刘魁立
Liu Kuili

2019 年 3 月
March 2019

安仁赶分社
Spring Equinox Fair in Anren

前 言
Preface

　　湖南省安仁县是湘东南的一个边陲小县。安仁之名，可溯源至《论语》：仁者安仁。昔始祖神农遍走安仁，"制耒耜奠农工基础，尝百草开医药先河"，被后人尊为"药王"。每逢3月春分时节，安仁人"择社日祭神以祈谷"，于香草坪处"日中而市"，进行躬耕祭祀、结社交易等传统民俗活动，并在开耕下田前喝下强身健体的汤药。这就是至今存续于湖南省安仁县的传统节气习俗活动——赶分社，它完好地保留了春分时祭神祈谷、服药开耕的传统习俗。赶分社期间进行大规模中草药交易的习俗至今也有1 000多年了，既有悠久而深厚的神农文化积淀，还有丰富的中药材资源和远近闻名的中草药交易，因而赶分社在民间又称"药王节"。

Anren is a small remote county in the southeast of Hunan Province. The name "Anren"(meaning "at ease with benevolence") is derived from a saying in the *Analects*, "The benevolent are at ease with benevolence". In the beginning of time, the footprints of the founding ancestor of the Chinese nation, Shennong, covered every inch of Anren, hand-making primitive ploughs, doing basic farm work, tasting herbs by the hundred and starting Chinese herbal medicine. He was revered as the "King of Herbal Medicine". At the Spring Equinox in the third month on the lunar calendar each year, Anren inhabitants agreed on a day for the fair and gave sacrificial offerings to Gods to pray for a bumper harvest and ran a market fair at the Spring Equinox at Xiangcaoping. Here, they farmed, prayed, celebrated, traded, conducted various traditional folk activities, and took a health-enhancing herbal medicine soup before starting to work in the fields. This is the Spring Equinox Fair—the traditional solar term folk celebration in Anren County, Hunan Province. It well preserves the traditional customs of presenting offerings to gods, praying for a bumper harvest, drinking herbal medicine and starting to plough, at the Spring Equinox. The custom of the large-scale trade of Chinese herbal medicine during the fair has been lasting for over 1,000 years. Because of the profound Shennong culture dating back to such a long time ago, the rich resources of Chinese herbal medicine, and the widely known trade of Chinese herbal medicine, the fair here is also popularly known as the Festival of the King of Medicine.

2016年11月30日，联合国教科文组织将"二十四节气"列入人类非物质文化遗产代表作名录，"安仁赶分社"作为十个非遗传承保护社区之一，被列入扩展名录。

On November 30, 2016, the UNESCO proclaimed the "Twenty-Four Solar Terms" a Masterpiece of Intangible Cultural Heritage of Humanity, and the "Spring Equinox Fair in Anren" was entered into the extended list of ten intangible heritage preservation communities.

人文地理

A Cultural and Geographical Overview

安仁赶分社节气习俗的传承地——湖南省安仁县，是全国生态魅力县，自古就有"桃源福地"之美誉。位于湖南省东南部，郴州市北端，东界茶陵县、炎陵县，南邻资兴市、永兴县，西连衡南县、耒阳市，北接衡东县、攸县，素有"八县通衢"之称。永乐江流贯全境。全县辖 5 镇 8 乡，总人口 46 万，面积 1 478 千米^2。其地处罗霄山脉西麓，地势东南高西北低，境内为半山半丘陵地区，山峻、水秀、洞奇。有"西

湖之美"之称的大源水库,有大石风景区,有"小桂林"美誉的赤滩
电站,有"云蒸雾海"的义海景区,有至今保持原始风貌的公木林场;
丹霞地貌范围广,面积大,连绵数里,有千冈、千峡、千湖、千坦之称;
更有全市出水量最大的温泉资源——龙海温泉,经勘探表明日出水可达
5万吨。

Anren County, Hunan Province, home to the heritage of the custom of Spring
Equinox Fair in Anren, is a National Ecologically Charming County, having
enjoyed a reputation of being "the origin of peach orchards and a land of
fortune". It is located in the southeast of Hunan Province, at the northern tip of
Chenzhou Municipality. Neighboring on Chaling and Yanling Counties in the
east, Zixing Municipality and Yongxing County in the south, Hengnan County
and Moyang Municipality in the west, and Hengdong and Youxian Counties
in the north, it has been known as the "hub of eight counties". The Yongle
River runs through the whole county. Five towns ("zhen") and eight townships
("xiang") are in its jurisdiction. The overall population is 460,000, and the
land area is 1,478km^2. The county is located at the western foot of the Luoxiao
Mountain Range, having a terrain of high ground in the southeast and lower
ground in the northwest, mostly semi-mountainous and semi-hilly, with towering
mountains, charming rivers and spectacular caves. Here, one can admire the
Dayuan Reservoir which is reputed to be "as beautiful as the West Lake", the
Dashi Scenic Zone, the Chitan Power Station known as "Miniature of Guilin",
and the Yihai Scenic Zone famous for its "steam-like clouds and sea-like fogs".
The ancient landscape has been preserved in Gongmu Forest Farm till today.
The Danxia Landform enjoys a wide range of topographic features, with a huge
expanse, continuing for a couple of kilometers, reputed as having one thousand
hills, one thousand gorges, one thousand lakes, and one thousand flat lands.
There is also the Longhai Hot Spring, a hot spring of the largest volume of spring
water in the municipality, with a daily water yield of as much as 50,000t.

永乐江国家湿地公园
Yonglejiang National Wetland Park

　　安仁县是湖南省有名的风景名胜区，也是全国三大最佳油菜花观赏地之一，素有"山川甲衡湘"的美誉。境内有享誉全国的国家 AAAA 级旅游景区——稻田公园、熊峰山国家森林公园、永乐江国家湿地公园、神农始祖殿等在内的"神农文化十大景区"。有机富硒米、豪峰茶、生态茶油等绿色农产品畅销全国。

Anren County is a famous scenic resort in Hunan Province, also one of the three best sightseeing spots for admiring rape flowers. It has always been reputed to have "best landscapes of mountains and rivers in Hunan". Within its jurisdiction, there are the "Ten Scenic Zones of Shennong Culture", including Daotian Park—the nationally reputed 4A-level tourist scenic zone, Xiongfengshan National Forest Park, Yonglejiang

National Wetland Park, and the Shennong Hall Dedicated to the Chinese Founding Ancestor. Its green agricultural products sell well in the whole country, including organic selenium-rich rice, Haofeng Tea, and ecological teaseed oil.

　　历史悠长、文化底蕴深厚的安仁县也素有"南国药都"之美誉。古有始祖神农"始尝百草，教化农耕"。轿顶屋是毛泽东和朱德井冈山会师的决策地，为国家级文物保护单位。安仁元宵米塑、安仁花鼓戏、安仁龙灯会都被列入湖南省级非物质文化遗产代表性名录，现有省、市级以上保护项目 26 个。有着深厚历史文化积淀的安仁，历史上也产生了如南宋名将韩京，文学家计有功、刘梦应、刘应祥，清代教育家欧阳厚均、清朝抗法民族英雄侯材骥等众多的名人雅士。

Apart from its long history and profound cultural accumulation, Anren County enjoys the reputation of "medical capital of South China". In ancient times, the founding ancestor Shennong first tasted and tested herbs by the hundred, and enlightened people on how to till and plough for farming. Jiaodingwu was the place where Mao Zedong and Zhu De decided to join their troops of the red army—

a national unit of preservation of cultural relics. Anren rice sculpture, Anren Huagu Opera and Anren Dragon Lantern Show have all been listed as Hunan provincial intangible cultural heritage. Anren has altogether 26 cultural protection units at provincial and municipal levels. With its profound historical and cultural accumulation, Anren produced numerous great personages, such as General Han Jing and great men of letters Ji Yougong, Liu Mengying, and Liu Yingxiang of the Southern Song Dynasty; educator Ouyang Houjun and anti-French-aggression national hero Hou Caiji of the Qing Dynasty.

清同治年间安仁县地图
Map of Anren County of Tongzhi
Era of Qing Dynasty

安仁县城
County Seat of Anren

历史渊源
Historical Origins

　　相传距今 4 700 多年前，中华农耕文明的创始者炎帝，率众在古荆州（今安仁境内）遍尝百草，日遇七十二毒，得荼（茶）而解之，遂得此灵感，拔草为药以医民恙。此外还发明农具耒耜以利耕耘、教民始种五谷以为民食、日中为市以利民生，可谓"功昭日月，德泽后世"。后炎帝因误食断肠草而亡。当地百姓为纪念炎帝"制耒耜奠农工基础，尝百草开医药之先河"的功德，在风光秀美的香草坪（今称安仁县永乐

江镇）兴建神农殿，供人们祭祀炎帝。又因中国人自古有着"择社日祭神以祈谷"的习俗，所以逐渐形成了在春分前后举行春社，以祭祀炎帝、祀神祈谷、开耕祈福以及开市交易。每当此时，来自五湖四海的商贩云集于此，捧售香烛、纸钱、草药、索、锄柄、斗笠以及各种农副土特产品。

Approximately 4,700 years ago, the founder of the Chinese agricultural civilization, Emperor Yan headed his followers in tasting and testing hundreds of herbs in the ancient Jingzhou (in what is now Anren County), and was poisoned 72 times within a day but was able to find an antidote in tea. Inspired by this, he uprooted the herbs as medication to cure people of their discomforts. He invented farm tools such as primitive hoes and ploughs for farming, taught people to sow five different kinds of grains for food, and arranged a marketplace at the beginning of the month to bring convenience to people, thus providing merits as bright as the sun and the moon, and exhibiting virtues that benefited all descendants for future generations. Subsequently, however, Emperor Yan ate the graceful jasmine herb by mistake and died as a result. In order to commemorate his merits of laying the foundation for farming and farm tools and launching medicine by tasting and testing hundreds of herbs, local inhabitants built the Shennong Hall in the scenic Xiangcaoping (currently called the Town of Yonglejiang, Anren County), where people could offer sacrifices to him. Meanwhile, the Chinese have had the custom of giving sacrifices to gods to pray for a bumper harvest at the Spring Equinox. Consequently, the vernal sacrifice around the Spring Equinox gradually evolved, to commemorate Emperor Yan, pray to gods for a bumper harvest, start the year's farming, ask for auspiciousness, and launch the marketplace for trading. At this time every year, traders and vendors all around would gather here, to sell incense, candles, paper money, herbal drugs, ropes, hoes and hoe handles, straw hats and various agricultural food and products.

安仁县这种独特的赶分社春分节庆习俗活动在五代后唐同光年间（923—926年）形成雏形。宋咸平五年（1002年），知县高岳徙县治于香草坪，定于每年春分节气社日，祭祀炎帝、开药市、开耕，逐渐形成了与今天并无二致的赶分社节气习俗。清《安仁县志》记载"择社日祭神以祈谷"，有"春分为期，香草坪为所，致天下之民，聚天下之货，交易而退，各得其所"之盛况。赶分社之节气习俗得以传承延续。

The embryonic form of this unique folk custom of visiting the fair and celebrating the Spring Equinox in Anren County evolved in the Tongguang Era (923—926) of the Later Tang Dynasty during the Five Dynasties period. In the fifth year (1002) of the Xianping Era of the Song Dynasty, the county magistrate Gao Yue moved the county seat to Xiangcaoping, and decreed that on the Sacrifice Day around the Spring Equinox every year, Emperor Yan would be given his offerings and prayers, the medical market launched, and farming started for the year. In this way, the custom of Spring Equinox Fair had gradually evolved which was very similar to how it is observed today. According to *The Annals of the Anren County*, the Sacrifice Day was chosen to commemorate gods and pray for a bumper harvest. It was quite a spectacle. At the time of the Spring Equinox, in Xiangcaoping, people from everywhere gathered here, goods from all around the country were gathered, and when they left after the transactions, all were content with the business concluded. The custom of the Spring Equinox Fair has thus been preserved and passed down till today.

神农殿祭祀
Offering Sacrifices at Shennong Hall

　　"赶分社"，究其含义，"赶"为动词，急赴之意，指无论是赶来祭祀的民众，还是赶来开市的商贩以及赶着开耕春播的农夫，都有不耽误、不错过、赶吉时之意。一则表示祭祀炎帝的感恩之心，二则表示祀神祈谷之迫切，三则表示赶在春分时不误农事，讨好彩头，四则表示开市大吉之寓意。社日是民间祭祀土地神的日子，一般在春天和秋天各举办一次，叫作"春社""秋社"。春社是在春分节气社日，"春祭社以祈膏雨，望五谷丰熟；秋祭社以百谷丰稔，所以报功"。安仁春分节赶分社就是在春分祭祀炎帝和祭土地，杀牛宰羊祭酒、开市交易、喝药酒、开耕。

The name of the custom in Chinese, "An Ren Gan Fen She", literally means rushing to participate in the sacrificial celebrations, referring to the speedy actions of the people who come here to participate in the sacrificial rites and the peddlers who are in a hurry to start business transactions here, and the farmers who are eager to start land tilling and spring sowing. None wants to miss the auspicious hour. First, they want to show their appreciation to Emperor Yan at his sacrificial rites; second, they are anxious to pray to gods for a bumper harvest; third, they want to participate in this event before the timely start of the farming activities, catch hold of the best luck for auspiciousness; and fourth, they hope the marketplace and fair will be smooth and successful. The Sacrifice Day is the day to present sacrificial offerings to gnomes (earthly deities), hosted once in spring and once in autumn, known as the Spring Sacrifice and Autumn Sacrifice respectively. The former is observed at the solar term of the Spring Equinox, when people pray for plenty of rains and a bumper harvest of all five grains. At the Autumn Sacrifice, the various crops are already ripe and a bumper harvest is reaped, and thus people show their appreciation. At the Spring Equinox Fair in Anren, people give sacrificial offerings to Shennong, known as Emperor Yan, and the earth deities. They butcher cows and sheep and present wine, launch the marketplace and business transactions, drink medicinal wines and start the year's tilling and plowing.

活动内容 Ritual Events

（一）祀神祈谷 Sacrifice to Deities to Pray for Bumper Harvests of Grains

社日当天，安仁县老百姓自发聚集在县城神农殿"祀社神以祈谷"。祭祀方式分为官方祭祀和民间祭祀两种，官方祭祀由政府官员主持，民间祭祀由道长主持。祭祀内容和流程为：鸣炮、上香、上贡、起乐、献谷草、供三牲（牛、羊、猪）、读祭文。上香、上贡、起乐、献谷草

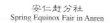
者皆为古代仕女装扮，以传统乐曲伴奏；祭文由当地最高行政长官宣读，祈求保佑"风调雨顺、五谷丰登，一方平安、人丁康乐"。祭拜礼成后，人们自由轮流祈愿。

On the very day of sacrifice, the local residents of Anren County spontaneously gather at the Shennong Hall of the county seat to pay tribute to the earth deities and pray for a bumper harvest. There are two different sacrificial ceremonies. The official rites are presided over by government officials and the folk rites are moderated by a Taoist priest. This is the flow process of the sacrificial ritual: sounding the cannons, offering incenses, presenting offerings, playing the music, presenting grains and grass, offering meat of three domestic animals (cow, sheep and pig), and reading the eulogy. Those who offer incense and sacrifice, play the music and present grains and grass are all dressed as maids in ancient China, to the accompaniment of traditional melodies. The eulogy is read by the top executive of the locality, praying for blessings with favorable winds and rains, bumper harvests of all five grains, peace and stability in the locality, and prosperity and happiness of the population. After the completion of the sacrificial rituals, the spectators take turns to offer their prayers for blessings.

神农殿祭祀
Offering Sacrifices at Shennong Hall

神农殿祭祀
Offering Sacrifices at Shennong Hall

祭文
Script of Eulogy

祭祀器乐演奏
Performance of Sacrificial Musical Instruments

（二）集会演出 Assemblies and Performances

春分社日期间，安仁县热闹非凡，从四面八方赶来的人们聚集于此，走亲访友，交流农事经验、养生之道。除了举行隆重盛大的祭社祀神仪式外，还有远近闻名的药市交易，各路民间艺人和戏班子的表演，如唱社戏、踩高跷、玩杂耍、走旱船、布袋皮影戏、龙狮舞等，真正是"你方唱罢我登场"，怎一个热闹了得。赶分社给人们提供了丰富的娱乐休闲活动，至夜晚于永乐江边放河灯，寄托对故人的思念和对未来美好生活的祈愿。其间，祁剧、湘剧、花鼓戏等各种地方戏种班子轮番上演，成为安仁县独具特色的春分社日民间文艺形式，其热闹狂欢程度媲美春节。

During the Spring Equinox and sacrificial celebrations, Anren County is spectacularly busy and noisy. People gather here from all around the county, visiting friends and relatives, exchanging agricultural and farming experiences as well as their regimen tips and tricks. Apart from the ceremonious and grand rituals for giving sacrifices to earth deities, there is also the marketplace for medicinal drug transactions known far and wide, and performances by various folk artists and opera groups. They sing sacrificial operas, walk on stilts, offer variety shows, row land boats, enjoy glove shadow puppetry, and watch dragon and lion dances. One performance is immediately followed by another and there is no pause in between. One is overwhelmed by all these different shows and performances. The event provides a rich variety of entertainment and fun activities. In the evening, people launch their river lanterns on the Yongle River, commemorating their deceased loved ones and giving best wishes for a beautiful life in future. During this period, various local operas take turns to give their performances: Qi Opera, Xiang Opera, and Huagu Opera and more. They make a unique and distinct folk art form on the Sacrifice Day at the Spring Equinox in Anren County. Its carnival-like boisterousness is comparable to the atmosphere during the Spring Festival.

祭祀现场龙舞表演
Dragon Dance Performance at Sacrificial Venue

祭祀现场腰鼓、采莲船表演
Waist Drum and Lotus Fruit Picking Boat Performances at Sacrificial Venue

旱船和蚌壳表演
Land Boat and Clamshell Performances

社戏表演
Temple Fair Performances

(三) 春分开耕 Start of Plowing at Spring Equinox

春分开耕分民间或县衙两种形式。最初开耕地定在南门洲旁的水田，现代开耕仪式基本上延续了古老的传统。

The start of plowing at the Spring Equinox could be conducted unofficially or at the county *yamen*. Originally, the venue for the start of plowing was designated at the paddy field beside Nanmenzhou. The modern ritual for the start of plowing is basically a continuation of the ancient ceremony.

开耕仪式上，十里八乡的百姓献上五谷、三牲供品，年长者朗诵祭文，祝愿风调雨顺、五谷丰登。负责开耕的人一般为长者或县衙官员，戴上斗笠蓑衣，撸起裤腿赤着脚，牵着一头脖围红绸的黄牛，掌犁开耕。肥沃的土地翻滚而起，标志着春耕春播的开始。民间在开耕之前，家家户户都要利用"赶分社"的时间，购买、交换或采集草药，回家按民间土方配药，放入陶罐炖猪脚熬成药膳，吃了能"驱寒壮骨、不怕水冷泥深、不易患风湿"，方可春耕。从2014年起，春分开耕仪式地定在稻田公园。

春分开耕
Launch of Plowing at Spring Equinox

春分开耕
Launch of Plowing at Spring Equinox

At the ceremony of the Start of Plowing, the masses from villages and towns all around presented their offerings of five kinds of grains and meat of three kinds of domesticated animals. A senior person reads aloud the eulogy, wishing for favorable winds and rains and bumper harvests of all five grains. The person who launches the start of plowing is usually someone senior in age or an official in the county *yamen*, who wears a bamboo hat and a coir raincoat, bare-footed with the trouser legs rolled up, drives a yellow cow with red silk cloths around its neck, and starts to plow the fields by using the plough. The fertile earth is churned, signifying the start of spring cultivation and sowing. Before villagers start their plowing, every household will purchase, trade or collect herbal medicines by taking advantage of the occasion of the Spring Equinox Fair. Back at home, based on folk formulas, they place these into pottery jars with pig trotters and boil these into medicinal meals. They are then eaten to keep out the cold and strengthen the bones, confront cold water and deep silt with better wherewithal, and resist rheumatism, and then the plowing can begin. Since 2014, the Daotian Park has been designated as the venue for this ritual.

（四）赶场交易 Hurrying to the Fair for Trade Deals

　　昔神农在安仁遍尝百草以医民恙，开创了中国传统中医学的先河，安仁也以其深厚的神农文化享誉湘江大地。安仁县自古是中草药的集散地，每年春分前后，商贾云集，来自四面八方的草药商人竞相交易，形成远近闻名的药市，至今已经有 1 000 多年的历史了，堪称中国历史上最早的药市。素有"药材不到安仁购不齐，药方不到安仁配不灵，郎中不到安仁诊不神"的说法。今天的安仁每逢春分社日，在县城南门洲上都会进行大规模的中草药材交易，还有大量的农耕工具及日常生活用品、土特产品的交易。

Originally, Shennong tasted and tested hundreds of herbs in Anren to cure people of their diseases, launching traditional Chinese medicine, and Anren County has been well famed throughout the Xiangjiang River basin for its profound Shennong culture. Anren County has been the hub of Chinese herbal medicines since ancient times. Around the Spring Equinox every year, dealers and peddlers gather here, and herbal medicine traders from everywhere vie with each other for transactions, and the medicine marketplace that has thus evolved is well known far and wide. With its history of over 1,000 years, the marketplace is reputed to be the earliest in Chinese history. There has been the saying that medicinal materials would be incomplete if the purchase is not made in Anren, prescriptions would not have the same curative effect if they are not given in Anren, and doctors would not be miraculously effective if they have not practiced in Anren. Today, at the Spring Equinox Day of Sacrifice every year in Anren, large-scale transactions of Chinese herbal medicinal materials are held in Nanmenzhou at the county seat, together with transactions in farming and plowing tools, daily life necessities, and local specialties from different regions.

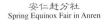
　　药材地摊琳琅满目，各种原生态的中草药一捆捆、一堆堆、一筐筐，堆积如山。有细如胡须的白草，有殷红如血的血葫芦，还有人形一样的何首乌……让人看得眼花缭乱。市场里吆喝声、询价声、嬉闹声和广告喇叭声连成一片，此起彼伏，堪称一绝。

There is an impressive array of stands for herbal medicinal materials. Various raw Chinese herbal medicines are packed in bundles and baskets and stacked in piles and hills. Discolored Cinquefoil Herbs are as thin as beard hairs, Shouliang Yam Tubers are as crimson as blood, and Tuber Fleeceflower Roots look like human figures. Well, they all jump into one's eyes, in a dazzling and perplexing manner. In the market, hawkings, loud inquiry exchanges, laughter and joking, and advertizing loudspeakers mingle with each other, which never seem to cease, a scene that cannot be matched anywhere else.

"赶分社"药市
Medicinal Market at Spring Equinox Fair

（一）神农殿 Shennong Hall

　　《安仁县志》记载，清康熙三十年（1691 年）知县陈黄永在县城北面凿泉井，修建洁爱亭一座，珠泉井一口，亭后山上构庵五楹。这就是安仁古八景之一的"泉亭珠涌"，北面山上为白衣庵、神农殿所在地。神农殿，是当地人们祭祀炎帝神农氏的地方。嘉庆年间知县周延瑾重修

庙宇，知县许潢重修泉亭。每年春分时节，人们在炎帝木主牌位前燃香草、香楮枝祭祀，并交易草药和耒耜。据史料记载，民国后期，白衣庵为一幢砖木结构的庵宇，坐北朝南。神农殿则坐西朝东，为一座方形古建筑，东面为一石拱门，五级台阶，堂上供奉炎帝的木雕塑像，供人们祭祀。殿周围建有四合院，两边天井内的两棵银杏树，每逢秋季到来，成熟的银杏果纷纷掉落地上，又成一景。

According to the *Annals of Anren County*, in the 30th year (1691) in the Kangxi Era of the Qing Dynasty, the county magistrate Chen Huangyong excavated a spring water well in the north of the county seat, and built the Jie'ai Pavilion with the Zhuquan Well in it. Behind the pavilion was a nunnery of five rooms in width. This was one of the eight ancient sights of Anren County, "pearl water welling under the Quanting Pavilion". On the hills in the north are located the Baiyi Nunnery and Shennong Hall. The Shennong Hall was where local people made sacrificial offerings to Emperor Yan, Shennong. During the Jiaqing Era of the Qing Dynasty, county magistrate Zhou Yanjin had the temple rebuilt, and county magistrate Xu Heng had the Quanting Pavilion reconstructed. Every year around the Spring Equinox, people burned fragrant grasses and fragrant paper mulberry twigs, and did transactions in herbal drugs and rakes and ploughs. According to historical documents, in the later half of the Republican Period, the Baiyi Nunnery was a temple of brick and wood structure facing south, while the Shennong Hall, an ancient square-shaped architecture facing the east, having an arch door in the east with five steps in front, had wooden sculptures of Emperor Yan enshrined here for people to worship. Around the hall was built a quadruple courtyard; ripe apricots fell on the ground around the two ginkgo trees in the patios on both sides of the courtyard each autumn, making one more spectacle in the area.

清同治年间神农殿位置图
Map of Shennong Hall in Tongzhi Era of Qing Dynasty

　　神农殿历经岁月沧桑，几经变迁。因为办学需求，曾先后作为女子职业学校、私立昆仑中学，1949 年后安仁县二中也曾设在这里。20世纪 70 年代神农殿旁的四合院被氮肥厂占用，西面、北面两栋平房随后被拆除。神农殿在几番征用之后，只剩下一处遗址。2001 年，县文物部门普查，在此发现几块"神农殿"字样的青砖。

The Shennong Hall went through many vicissitudes and changes. For educational needs, it was used as the Vocational School for Women, and the Private Kunlun Middle School, as well as the No. 2 Anren County Middle School after 1949. In the 1970s, the quadruple courtyard beside the Shennong Hall was occupied by the Nitrogen Fertilizer Factory, and two bungalows in the west and in the north were subsequently dismantled. After these rounds of expropriation, the Shennong Hall became only a relic site. In 2001, the Department of Cultural Relics of the County discovered several blue bricks with characters of "Shennong Hall" while conducting a general survey.

神农殿遗址发现的青砖
Blue Brick Excavated at the Relic Site of Shennong Hall

2000 年，为弘扬神农文化，通过民间筹资、政府扶助，在县城东南凤冈山上新建了一座神农殿，占地 12 万米^2，分四个阶梯广场，共有 199 级台阶拾级而上，殿高 20.8 米、长 42.8 米、宽 32.8 米，整个建筑采用仿古建筑风格，内有一座由安仁籍著名陶艺家周国桢先生设计的 7 米高的炎帝塑像。塑像手持药锄、足踏草鞋，座前摆放一个竹制药篓子，其装束神情与一位准备上山挖药草的老农极其相似。2004 年 3 月，安仁县人民政府举行"中国·安仁春分药王节暨神农殿落成典礼"。从此，安仁赶分社，人们祭祀炎帝有了一个新的处所，实现了民众的一个夙愿。

In 2000, in order to promote the Shennong culture, with private finance and government support, a new Shennong Hall was built on the Fenggang Mountain in the southeast of the County seat, occupying an area of 120,000m^2, equipped with four terraced plazas, with 199 steps leading up to the hall; the hall is 20.8m tall, 42.8m long and 32.8m wide. The whole structure adopts an archaized architectural style, with a seven-meter-tall sculpture of Emperor Yan enshrined inside, designed by the reputable sculptor Zhou Guozhen, originally a native of Anren. The sculpture has a herbal medicinal hoe in its hand, wears a pair of straw sandals, and has a bamboo basket for medicinal herbs in front of the pedestal, its apparel and overall look closely resembling an elderly farmer who is ready to go into the hills to pick or dig for medicinal herbs. In March 2004, the People's Government of Anren County hosted the "Anren Spring Equinox Festival of King of Medicine and the Launching Ceremony of Shennong Hall". From then on, when people vie with each other to participate in the Spring Equinox Fair in Anren, they have a new venue for commemorating and presenting their sacrificial offerings to Emperor Yan, thus realizing a long-pursued dream of the masses.

神农殿
Shennong Hall

（二）神农广场 Shennong Square

神农广场修建于 2013 年，占地 5.3 公顷。广场布局暗含太极八卦，正中央为炎帝全身雕像，高 9.5 米，象征着九五之尊；底座宽 4.26 米，代表神农的生日。炎帝头生牛角，执耒耜，捧灵芝，缀令牌，脚踏风霜，面带坚毅，目含慈悲，寓意其"始作耒耜，教民农耕，遍尝百草，发明医药，日中为市，首倡交易"等千秋功绩。华佗、李时珍、孙思邈等八位古代医圣环伺八方，狮、龙、象等 108 座神兽拱卫四周，另有图腾柱 4 根。广场与神农殿处在同一中轴线，以立交桥相连，浑然一体，是市民休闲娱乐、民众祭拜炎帝的一个重要活动区域。

The Shennong Square was built in 2013, occupying 5.3ha of land. The layout of the square conforms implicitly with the *Tai Chi* ("supreme ultimate") and *Ba Gua* ("eight trigram palm"). In the center is the full-length sculpture of Emperor Yan of a height of 9.5m, symbolizing his dignity of a sovereign (out of the nine single digits, "nine" means supreme and "five" means medium and humble). The pedestal is 4.26m wide, indicating Shennong's birthday (April 26). The sculpture has a pair of bull horns on its head, carries a rake in one hand, holds a *lingzhi* mushroom (glossy ganoderma) in the other, wears a security token for decree-giving occasions, rides on wind and frost, wears a stern look of fortitude, and beams with mercy and benevolence in its eyes. All this is to appreciate Emperor Yan's timeless contributions in first producing rakes and ploughs, showing people how to farm and carry out agriculture, tasting and testing hundreds of medicinal herbs, inventing medical drugs, launching the marketplace at the Spring Equinox, and promoting business transactions in the first place. Sculptures of eight ancient medical sages, including those of Hua Tuo (145—208), Sun Simiao (541—682), and Li Shizhen (1518—1593), stands around Emperor Yan in attendance; 108 sculptures of divine animals, including lions,

dragons and elephants, are in postures and positions ready to defend Emperor Yan; there are also four totem columns. The square and the Shennong Hall are located along the same central axis, connected by a flyover and thus forming an organic whole, serving as an important venue and zone for local residents to have fun and entertain, and present sacrificial offerings to Emperor Yan.

神农广场
Shennong Square

神农广场
Shennong Square

（三）神农文塔 Literati Pagoda of Shennong

神农文塔位于安仁县熊峰山国家森林公园内的熊峰山顶，塔高七层，底层设有塔座及文化碑廊。相传，炎帝神农氏在安仁遍尝百草、辨识药性、传授医术、开创药市、试行农耕，在熊峰山教化山民结绳累积以计数，依形画样而记事。神农氏在安仁豪山误食断肠草而亡之后，未被抬回熊峰山一带安葬，山民在他时常逗留的熊峰山顶垒石成堆，不时跪拜，以不忘神农之功德。久而久之，石堆渐变为石塔，人称神农文塔。慢慢地，神农文塔便成为熊峰山一带人们祭祀神农、祈承文脉的一个神圣处所。

The Literati Pagoda of Shennong is located on top of the Xiongfeng Mountain, inside the Xiongfeng Mountain National Forest Park in Anren County. The pagoda is seven storeys tall, and on its ground storey are designated the pedestal and the cultural porch of steles. The legend has it that Emperor Yan, Shennong, tasted and tested hundreds of herbs meticulously, identified their medicinal qualities, enlightened people on medicine, launched pharmaceutical markets, and attempted farming and plowing, all over Anren. He also showed the inhabitants of Xiongfeng Mountain to calculate by tying knots with ropes, and to record happenings by drawing pictures. After Shennong died from taking the graceful jasmine herb by mistake in Haoshan Mountain in Anren, he was carried back to Xiongfeng Mountain for burial. Thus, local inhabitants of Xiongfeng Mountain piled up some rocks where he had often lingered, and knelt and kowtowed in front of the pile of rocks to commemorate his virtues and contribution. Gradually, the pile of rocks became a pagoda, known as the Literati Pagoda of Shennong. In time, the Pagoda became a holy shrine for people around Xiongfeng Mountain to pay sacrificial tribute to Shennong and to pray for an endless succession of literati.

因为神农的庇佑，安仁人文蔚起。从唐至清，共中进士28人，举人107人，贡生366人。南宋咸淳四年（1268年）戊辰科，衡郡中进士6人，其中就有4个安仁人。自古以来，安仁人物中，廉能兼著，或出仕，或从戎，或讲学；著书立说、洁己爱民、乐善好施者，大有人在。安仁县民间流传这样的说法：不拜神农文塔者，其子弟难以承继书香之气。

With Shennong's patronage, the number of great literati has been impressive. From the Tang Dynasty to Qing Dynasty, there were 28 palace graduates, 107 provincial graduates, and 366 senior licentiates (or known as tribute students). In the fourth year (1268) in the Xianheng Era of the Southern Song Dynasty, six people from Hengjun Prefecture became palace graduates, four of whom were from Anren. Since ancient times, among famed personages from Anren, of great virtue and with great literary talents, some became government officials, some joined the military forces, and some became instructors. The number of people who wrote treatises and launched their own literary theories, who were incorruptible and showed great love for the people in their jurisdictions, and who were philanthropic and charitable, was impressively large. In Anren County, a saying goes like this: without praying to Shennong at the Literati Pagoda of Shennong, one's children would not be likely to inherit and develop the cultural and literary tradition of the family.

神农文塔初建于明万历年间，清代乾隆年间重修。清嘉庆二十三年（1818年），得到捐资再次修缮。不久，进士卢兆鳌和士绅林添瑞增建二塔于神农文塔左右供民众参拜，俗称"三柱塔"。

The Pagoda was originally built in the Wanli Era of the Ming Dynasty, and renovated in the Qianlong Era in the Qing Dynasty. In the 23rd year (1818) of the Jiaqing Era in the Qing Dynasty, adequate donations had been raised for another renovation. Soon afterwards, senior licentiate Lu Zhao'ao and Squire Lin Tianrui built two more pagodas on both sides of the Literati Pagoda of Shennong as

supplementary shrines for the pious believers. They have been popularly known as the "three-column pagoda".

神农文塔
Literati Pagoda of Shennong

清同治年间神农文塔位置图
Map of Literati Pagoda of Shennong in Tongzhi Era of Qing Dynasty

（四）神农文化街 Shennong Culture Street

神农文化街位于县城东面凤冈山脚下，与神农殿、神农广场相邻，始建于 1993 年。街长 1 000 多米，宽 20 余米，街道两端建有仿古牌楼，临街店铺和走廊都依徽派建筑风格打造，青砖青瓦，飞檐画栋，古色古香。这里主要是中草药交易、加工和民间郎中坐诊的地方。每年春分，也是四方宾客相约安仁赶分社必去之地。

The Shennong Culture Street is located at the foot of the Fenggang Mountain in the east of the county seat, adjacent to the Shennong Hall and the Shennong Square. First built in 1993, it is more than 1,000m long, over 20m wide, with archaized pailoos at both ends of the street. The shops on both sides of the street and the corridors adopt the Anhui architectural styles—blue bricks and tiles, flying eaves and painted walls, beaming with a tint of ancient styles. This is the major venue for transacting and processing Chinese herbal medicines, and also for private doctors to give consultation to patients. At the Spring Equinox each year, visitors and guests will definitely visit the place when they vie with each other to attend the Spring Equinox Fair.

神农文化街
Shennong Culture Street

　　2015 年，为加大对安仁赶分社这一千年民俗活动项目的保护，安仁县委、县政府对神农文化街进行了修缮，使文化街更具实用性、民俗性、观赏性、艺术性，成为一道亮丽的神农文化风景线。

In 2015, in order to enhance the protection of Spring Equinox Fair in Anren, the folk culture event with a history of 1,000 years, the Anren County Party Committee and the Anren County Government sponsored renovation of the Shennong Cultural Street, and so it has become more practical and artistic, more oriented in folk cultures and worthy of viewing, making an impressive array of the Shennong culture scenery.

（五）神农百草园
Shennong Garden of a Hundred Herbs

神农百草园坐落于洋际乡茅坪村，距安仁县城 7 千米。园区占地
200 公顷，包括木本、草本、灌木和菌类四大区域，大面积种植了枳壳、
丹参、白术、栀子等 100 多种中药材。神农百草园的创建，旨在颂扬
神农氏为民奉献、不畏艰险、身体力行、勇于创新的精神，是倡导绿色
生态理念的实践，也是对安仁赶分社自然中草药不足的一种弥补。

The Shennong Garden of A Hundred Herbs is situated at the Maoping Village, Yangji Township, 7km from the seat of Anren County. The Garden occupies an area of 200ha, divided into wood, grass, shrub, and fungus zones, with more than 100 Chinese medicinal herbs planted in large areas, including bitter oranges, red sage roots, atractylis ovatas, and gardenias. The garden is built to promote Shennong's spirit of devotion, selflessness, hands-on practice and innovation. It is an application and support of the concept of green ecology, and also a compensation for the insufficiency of the natural Chinese medicinal herbs during the Spring Equinox Fair in Anren.

神农百草园
Shennong Garden of a Hundred Herbs

（六）中草药市场——石头坝市场
The Chinese Medicinal Herbs Market – Shitouba Marketplace

据史料记载，安仁赶分社中草药交易，最早是在县城（香草坪）老街边的南门洲上，起初主要为草药、耕牛、谷种、农具等生产生活用品交易，也伴随有相亲走访、农事交流等活动。后因城市规划发展需要，赶分社中草药交易场地几经变迁。2004 年石头坝中草药市场建成并运营后，安仁赶分社中草药交易终于有了相对固定的场所。每值春分，中草药交易额上亿元。

According to historical accounts, the transactions of Chinese herbal medicines during the Spring Equinox Fair in Anren was originally held at Nanmenzhou, to the south of the old street of the county seat (Xiangcaoping), for trading articles of daily life and production, such as herbal medicines, plowing cows, grain seeds, and farm tools. It was also an occasion of visits and dates, as well as exchanges of farming experiences. Later on, due to the requirements of city planning and development, the venue for the transactions of Chinese herbal medicines during the Spring Equinox Fair has been changed a number of times. When the Shitouba Chinese Herbal Medicine Market was built and launched into operation in 2004, the venue for the transactions of the Chine herbal medicines was finally fixed. At each Spring Equinox, the volume of the transactions in Chinese herbal medicine runs as much as RMB 100 million yuan.

石头坝中草药交易市场
Shitouba Marketplace of Chinese Medicinal Herbs

（七）药湖寺 Yaohu Temple

药湖寺位于安平镇药湖村，距安仁县城32千米，是炎帝洗药的地方。据传，炎帝率8名随从采完药后，在一水塘边洗药，并用采来的草药为民治病。时间久了，洗过百草的潭水药香四溢，炎帝笑对随从说："此塘真乃药湖也"。后人为感念炎帝的功德，在古碧海兴建殿宇，取名为药湖寺。据史料证实，北宋、南宋、元朝、明朝、清朝等历朝历代都非常重视药湖寺，当政者多次拨资修缮。

The Yaohu Temple is located at the Yaohu Village in the town of Anping, 32km from the county seat of Anren. It was where Emperor Yan washed and cleaned herbal medicines. According to the legend, after Emperor Yan had picked the herbs with his eight attendants, he would cleanse the herbs at the pond and provide treatment for local patients with his newly picked herbal medicines. Gradually, the water of the pond was filled with medicinal fragrance. Emperor Yan smilingly commented to his attendants, "This pond is indeed a medicinal lake". In order to show appreciation to Emperor Yan's virtues and merits, a temple was built at the Gubihai Lake, and named it Yaohu Temple (Temple of the Medicinal Lake). According to historical documents, the Yaohu Temple was given much attention in later dynasties, including Northern Song, Southern Song, Yuan, Ming and Qing Dynasties. Governors and magistrates appropriated funds for its repairs and renovation, on many occasions.

（八）稻田公园——二十四节气公园
Daotian Park – A Park Dedicated to the Twenty-Four Solar Terms

安仁县稻田公园位于永乐江镇，是一座集农业示范、农耕体验、科普教育、旅游观光、休闲娱乐于一体的二十四节气公园和农业湿地公园。园名由中国科学院院士、"杂交水稻之父"袁隆平先生亲笔题写。整个园区 20 千米^2，花、草、树、景等公园元素恰到好处地融入 3 333 公顷连片稻田中，与永乐江、神农景区、熊峰山国家森林公园相互映衬，园内生态庄园、稻香村、农耕博物馆与山、水、田园浑然天成，相得益彰。

The Daotian Park of Anren County, located in the town of Yonglejiang, is a park dedicated to the twenty-four solar terms and one of agricultural wetland, providing agricultural demonstration, letting visitors experience farming and plowing, carrying out science popularization and education, facilitating travelling and tourism, and offering leisure and entertainment. The name of the park was written by Professor Yuan Longping, academician of the Chinese Academy of Sciences and "father of hybrid rice". The park occupies a land of 20km^2, where such elements of the park as flowers, grasses, trees and scenery are beautifully and aptly integrated into the whole

stretch of the rice field of 3,333ha. It is well coordinated and matched with Yongle River, Shennong Scenic Zone, and Xiongfeng Mountain National Forest Park. Within the park, the ecological manor, the Daoxiang Village, and the Agricultural Cultivation Museum are organically integrated with the mountains, the water bodies and the fields. Each element brings out the best of the others.

稻田公园——二十四节气公园
Daotian Park—A Park Dedicated to the Twenty-Four Solar Terms

传承保护
Preservation and Protection

　　为了让湖南郴州地区本土的春分节气民俗活动——"安仁赶分社"这一人类非物质文化遗产保护项目得到传承和保护，在上级主管部门和业务对口单位精心指导下，在安仁县委、县政府的领导下，安仁县开展了一系列传承保护工作。

For better preservation and protection of this intangible cultural heritage of humanity, Spring Equinox Fair in Anren, the local folk custom event at the solar term of Spring Equinox in Chenzhou, Hunan Province, under the meticulous and

close guidance of the superintendent responsible departments and counterpart departments, and under the leadership of the Anren County Party Committee and the County Government, Anren County has carried out a whole series of preservation and protection endeavors.

（一）组织普查调研
Large-Scale Surveys and Field Research

组织人员走访调查，搜集整理中草药 700 多味，整理编辑了《中草药七百味》一书，完成《安仁赶分社》50 米画卷创作。

Field visits and surveys have been conducted, and consequently more than 700 herbal medicines were collected and recorded, the book *700 Chinese Herbal Medicines* was compiled and edited, and the 50m-long painting scroll *Spring Equinox Fair in Anren* was completed.

（二）培训项目传承人 Training of Heritage Inheritors

2017 年举办了皮影戏传承人、龙狮传承人、旱船传承人、米塑传承人培训等项目，共培训人员达 100 人。开展民俗活动，组织举行神农殿祭祀、开耕仪式、龙狮表演、皮影戏表演、社戏表演、米塑表演等传统春分节气民俗文化活动。

In 2017, a number of inheritor training programs were sponsored, for the shadow puppet show, the dragon and lion dance, the land boat dance, and rice sculpture, with 100 personnel received training. Traditional folk custom events at the Spring Equinox were staged: Sacrificial Rites at Shennong Hall, Ceremony for Launch of Plowing, dragon and lion dance shows, shadow puppet shows, theatrical performances at the Equinox, and rice sculpture performances.

开展旱船表演培训
Land Boat Dance Performance Training

开展米塑技艺培训
Rice Sculpture Skills Training

开展龙狮表演培训
Dragon and Lion Dance Performance Training

开展皮影表演培训
Shadow Puppet Show Performance Training

（三）宣传展示活动 Publicity and Promotion

为了传承和保护、宣传和展示安仁赶分社，开展了一系列活动。举办安仁赶分社传承与保护工作座谈会，制作安仁赶分社宣传片，举办文化与自然遗产日活动，邀约媒体报道相关活动。2017 年，有 50 多家媒体对安仁赶分社的开幕式及祭祀和开耕仪式、传承与保护工作座谈会、皮影戏等民俗表演项目进行了全程报道，吸引了近 30 万外来游客，提高了安仁赶分社的文化影响力，进一步带动了安仁文化旅游经济的发展。

A series of activities and events have been staged in order to preserve, protect, promote and publicize the Spring Equinox Fair in Anren. A working consultation meeting was held on the theme of its preservation and protection, a publicity documentary was produced, and an event was sponsored on the Cultural and Natural Heritage Day. Media were invited for news coverage. In 2017, more than 50 media conducted full coverage of the opening ceremony of the Spring Equinox Fair in Anren, sacrificial rites, ceremonies of launch of plowing, the work seminar on preservation and protection, and the folk custom performances including the shadow puppet shows. Over 300,000 tourists arrived. Consequently, the cultural impact of the Spring Equinox Fair in Anren has been heightened, and the cultural and tourist economy of Anren has been further promoted and developed.

安仁赶分社传承与保护工作座谈会
Work Seminar on Protection and Preservation of Spring Equinox Fair at Anren

雨中开耕
Launch of Plowing in Rain

米塑制作表演
Rice Sculpturing Show

雨中开耕
Launch of Plowing in Rain

（四）征集实物 Requisition of Artifacts

征集了皮影戏演出设备一套，水车、石磨、风车、陶器制作轮等近 20 件农耕器物。

A complete set of performance equipment of the shadow puppet show has been requisitioned, together with 20 artifacts for farming and cultivation, including a water wheel, a stone grinder, a wind mill, and a porcelain production wheel.

图书在版编目（CIP）数据

安仁赶分社／中国农业博物馆组编 ．—北京：中国农业出版社，2019.5

（"人类非物质文化遗产代表作——二十四节气"科普丛书）

ISBN 978-7-109-25400-8

Ⅰ．①安… Ⅱ．①中… Ⅲ．①二十四节气-风俗习惯-安仁县-通俗读物 Ⅳ．① P462-49 ② K892.18-49

中国版本图书馆 CIP 数据核字（2019）第 060181 号

中国农业出版社出版

（北京市朝阳区麦子店街 18 号楼）

（邮政编码 100125）

责任编辑　张德君　李　晶　司雪飞

文字编辑　杨　春

————————————

北京中科印刷有限公司印刷　　新华书店北京发行所发行

2019 年 6 月第 1 版　　2019 年 6 月北京第 1 次印刷

————————————

开本：787mm×1092mm　1/16　印张：4

字数：80 千字

定价：40.00 元

（凡本版图书出现印刷、装订错误，请向出版社发行部调换）